Brice Assimizele

Optimization of Wood Fuel Chips Production in Denmark

Brice Assimizele

Optimization of Wood Fuel Chips Production in Denmark

LAP LAMBERT Academic Publishing

Impressum / Imprint

Bibliografische Information der Deutschen Nationalbibliothek: Die Deutsche Nationalbibliothek verzeichnet diese Publikation in der Deutschen Nationalbibliografie; detaillierte bibliografische Daten sind im Internet über http://dnb.d-nb.de abrufbar.

Alle in diesem Buch genannten Marken und Produktnamen unterliegen warenzeichen-, marken- oder patentrechtlichem Schutz bzw. sind Warenzeichen oder eingetragene Warenzeichen der jeweiligen Inhaber. Die Wiedergabe von Marken, Produktnamen, Gebrauchsnamen, Handelsnamen, Warenbezeichnungen u.s.w. in diesem Werk berechtigt auch ohne besondere Kennzeichnung nicht zu der Annahme, dass solche Namen im Sinne der Warenzeichen- und Markenschutzgesetzgebung als frei zu betrachten wären und daher von jedermann benutzt werden dürften.

Bibliographic information published by the Deutsche Nationalbibliothek: The Deutsche Nationalbibliothek lists this publication in the Deutsche Nationalbibliografie; detailed bibliographic data are available in the Internet at http://dnb.d-nb.de.

Any brand names and product names mentioned in this book are subject to trademark, brand or patent protection and are trademarks or registered trademarks of their respective holders. The use of brand names, product names, common names, trade names, product descriptions etc. even without a particular marking in this works is in no way to be construed to mean that such names may be regarded as unrestricted in respect of trademark and brand protection legislation and could thus be used by anyone.

Coverbild / Cover image: www.ingimage.com

Verlag / Publisher:
LAP LAMBERT Academic Publishing
ist ein Imprint der / is a trademark of
AV Akademikerverlag GmbH & Co. KG
Heinrich-Böcking-Str. 6-8, 66121 Saarbrücken, Deutschland / Germany
Email: info@lap-publishing.com

Herstellung: siehe letzte Seite /
Printed at: see last page
ISBN: 978-3-659-38901-6

Zugl. / Approved by: Molde, Molde University College, A specialized University in Logistics, Diss., 2012

TABLE OF CONTENTS

Chapter 1 Introduction..3

Chapter 2 Challenges faced by the supplier of woods chips 5

   2.1 Problem description...5

   2.2 Type of problem 6

   2.4 Research questions and objectives...7

Chapter 3 Literature Review 8

   3.1 Supply Chain Management (SCM) 8

   3.2 Vendor Manage Inventory (VMI) 8

   3.3 Production, Inventory, Distribution, Routing Problem (PIDRP) ...9

   3.4 Inventory Routing Problems (IRP) 10

      3.4.1 Deterministic Inventory Routing Problems (DIRP) 11

      3.4.2 Stochastic Inventory Routing Problems (SIRP)....................11

   3.5 Generalized Traveling Salesman Problem (GTSP) 11

Chapter 4 Description of the wood chips supply chain 13

   4.1 Production 16

      4.1.1 Chipping at roadside..16

      4.1.2 Chipping at the terminal 17

      4.1.3 Chipping at the plant 17

   4.2 Transportation 18

   4.3 Consumption 18

   4.4 A mip model for the pidrp..21

      4.4.1 Parameters and variables...22

      4.4.2 The Model...23

      4.4.3 Model description..24

Chapter 5   Solution Methods 26

   5.1 Exacts Methods 26

   5.2 Heuristics Methods 26

Chapter 6 Problem Decomposition 27

   6.1 Scope and Specifics objectives 27

   6.2 A MIP model for the Yearly Production Plan.................................29

      6.2.1 Parameters and variables...29

      6.2.2 The Model..30

6.2.3 Model Description 31

6.3 A tabu search based metaheuristic for intra districts routing.....................32

6.3.1 Solution representation..33

6.3.2 Moves and neighborhood 34

6.3.3 Initial solution...35

6.3.4 Tabu list and aspiration criteria 35

6.3.5 Proposed algorithm...36

6.4 Implementation 37

6.5 Data collection 38

Chapter 7 Computational experiments..40

7.1 Test cases...40

7.1.1 Test cases for small size instances..40

7.1.2 Test case for large size instances..42

7.1.3 Test case for real-world instances...43

7.2 Results and analysis 43

7.1.1 Test results and analysis for small size instances 43

7.1.2 Test results and analysis for large size instances 53

7.1.3 Test results and analysis for real-world instances 53

Chapter 8 Conclusions and future research 54

References..56

Appendix : Real-world data 62

CHAPTER 1 INTRODUCTION

Globalization and advancement in technology have increased competition between logistics industries. As a result, logistics systems or supply chain networks should provide excellent customer service by fulfilling the six "rights": ensuring that the right goods, in the right quantities, in the right condition, are delivered to the right place, at the right time, for the right cost. Logistics companies have to improve service quality, reduce price and lead time in order to survive and succeed in this new business environment where the customer is a powerful king. That is, the competitive edge is gained through the cost-effective fulfillment of orders. Operations research applied to supply chain management problems is then an important and critical tool or essential decisions support assistant to logistics and supply chain managers.

The book makes use of operations research to optimize the production and distribution of woods chips with Forest and Landscape Denmark (FLD) which is an independent research centre at the University of Copenhagen. FLD is focused on research, education and consultancy services in the area of forest, landscape and planning. The research center has about 300 employees, seven locations throughout Denmark with an annual budget of about 20 million Euros.

FLD was established on 1 January 2004 from a merger of the Danish Forest and Landscape Research Institute, the Danish Forestry College, the Danida Forest Seed Centre and parts of the Department of Economics and Natural Resources of the former Royal Veterinary and Agricultural University of Denmark (KVL). The purpose of the merger was to strengthen the professional environment and to create a unit with strong international capacities and experiences that are relevant to the development and environment program. It has as a mission to contribute to the increased welfare of present and coming generations. This is through improved rural and urban planning, and sustainable management and utilization of trees, forests, landscapes and other natural resources.

In the field of forest, landscape and planning, FLD work on:

- ➢ Education and training
- ➢ Development and environmental assistance
- ➢ Monitoring
- ➢ Consultancy services and extension
- ➢ Ministerial services
- ➢ Research and development (Danish Energy Authority, 2002).

Our task, in collaboration with FLD, was related to the last bullet; the optimization of the production and distribution of woods chips.

The rest of the book is divided as follows. The research problem is presented in chapter 2, followed by the literature review in chapter 3. We describe the wood chips supply chain in chapter 4 and present the research design in chapter 5. The developed mathematical model and a tabu search based meth heuristic methods are described in chapter 6. Chapter 7 presents the computational experiments of the proposed model and algorithm. Finally, the conclusion and future research are presented in chapter 8.

2.1 PROBLEM DESCRIPTION

The energy policy in Denmark is in constant change, mostly due to the environmental concern. One of the objectives of its energy plans is to increase the consumption of renewable energy by 100%. Thus, the increased use of straw and wood chips at centralized power plants is an important initiative in the field of biomass. Based on the biomass agreement, the power plants should use a certain amount of woods chips for energy. For instance, the power plants were supposed to use 200,000 tonnes of wood chips per year, which is equal to approx. 250,000 m^3 s. vol, as for 2004 agreement. The wood chips used by the power plants are produced from the Danish forest.

The supplier produces woods chips from felled trees in the forest. The Danish forest is subdivided into different districts and each district is in turn composed with different stands. For that reason, the chipper which produces wood chips has to move to different districts and move from stand to stand inside a given district in order to produce wood chips. The produced chips are then transported directly to customers (power and heating plants) or to the storage terminal as inventory. Figure 2.1 presents a simplified description of the production and distribution of wood chips. The chipper uses the route presented in black color to produce woods chips from felled trees at different stands located in different districts. Then the produced woods chips are transported to the terminal storage or to power plants by full load trucks. A detailed description of the supply chain is presented in chapter 4.

The production and distribution of fuel chips is very complex and difficult mostly because of the weather condition and the high production cost. In fact, the production cost is expensive normally because of the high relocation cost of the chipper between districts and stands as illustrated in figure 2.1. The production capacity is highly reduced in winter time because of the weather condition.

5

The supplier has to meet the demand from power and heating plants regardless of the weather condition, machine breakdowns etc. In addition, the storage capacities at the customers are limited. This leads to a considerable mismatch between the production and the consumption of wood chips.

In order to cope with the unbalanced supply and demand, 20% of the transported wood chips to heating and power plants are stored for a period of time at the storage terminal.

However, the storage cost is very high because of the extra handling cost, the dry matter loss during storage and the high capital cost.

We propose solutions that will help the supplier to meet the demand at the minimum production, distribution and inventory costs.

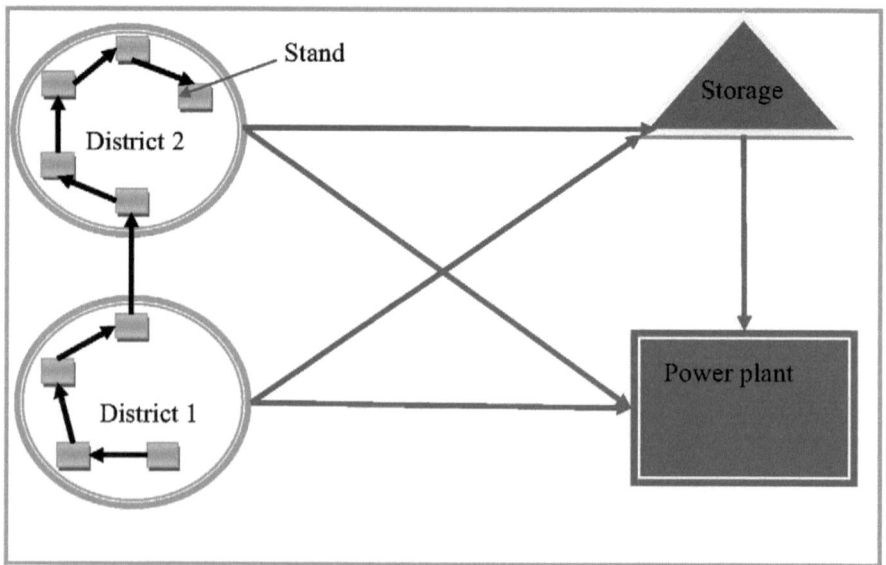

FIGURE 2.1: AN ILLUSTRATION OF THE WOOD

2.2 TYPE OF PROBLEM

The inventories of fuel chips (wood chips) at power and heating plants are managed by the supplier. That is, the vendor has to make sure there is a sufficient amount of

6

fuel chips at each plant. Clearly, the supply chain strategy used here is the Vendor Managed Inventory (VMI). This type of problem is defined in the operations research literature as Inventory Routing Problem (Aghezzaf, Raa, and Van, 2006) which will be described in the literature review section. The problem is extended to a Production, Inventory, Distribution, Routing Problem (Bard and Nananukul, 2010), as the objective is to coordinate the different component of the supply chain. Indeed, we consider not only the distribution of woods chips to customers but mainly the production as well.

2.4 RESEARCH QUESTIONS AND OBJECTIVES

The aim of the book is to suggest answers to some of the following questions:

1. How many fuel chips to produce each week?
2. As the forest is divided into districts, what is the best production sequence?
3. As the production rate and volume is considerably different at each stand, which stand and consequently, district should be visited for production in a given week and day?
4. When to visit each power plant for inventory replenishment?
5. How much to deliver to a power plant during a visit?
6. Which delivery routes to use?
7. How much inventory should be stored at the terminal each week?

In order to cost-effectively answer the above questions, we proposed the following research objectives:

➢ Describe and identify the critical parts of the supply chain
➢ Develop deterministic mathematical models for the problem
➢ Solve the problem by means of exact and heuristic methods

Specific objectives will be presented later in chapter 6 of this document.

CHAPTER 3 LITERATURE REVIEW

3.1 SUPPLY CHAIN MANAGEMENT (SCM)

Logistics management has been recognized, only in recent past, as a very important key in achieving competitive advantage. It can be defined as the process of managing strategically the procurement, movement of materials and storage as well as the related information flow through the organization and its marketing channels in a way that will maximize the current and future profitability through the cost-effective fulfillment of orders (Christopher 2011).

The logistics framework aims to create a single plan for products and information flow through the business (Harrison and Hoek 2011; Christopher 2011) while supply chain management has as objective to build on that framework, link and coordinate between the processes of the suppliers, customers and the organization itself.

It focuses on co-operation, trust and on the idea that properly managed, the 'whole can be greater than the sum of its parts' (Christopher 2011). Indeed, individual objectives often have conflicts in the entire supply chain (Hugos 2011). In that direction, SCM can be defined as: "the management of upstream and downstream relationships with suppliers and customers in order to deliver superior customer value at less cost to the supply chain as a whole" (Christopher 2011). It is then important to look at the optimization of all or important supply chain activities as a whole. For our case, the main collaborative strategy between the buyer and supplier to achieve this goal is the vendor manage inventory (VMI).

3.2 VENDOR MANAGE INVENTORY (VMI)

Vendor manage inventory is a retailer-vendor relationship where the vendor decides on the appropriate inventory levels within bounds that are agreed by contract with the retailer. Usually, the vendor incurs a penalty cost for items exceeding these bounds (Darwish and Odah 2010). The replenishment control is placed at the vendor who is then allowed to choose the timing and size of deliveries (Andersson et al. 2010).

VMI is a pull replenishing practice that is set up to allow a Quick Response (QR) from the supplier to actual demand. It represents the highest level of partnership where the order placement and inventory control decisions are totally controlled by the vendor (Tyan and Wee 2003).

Companies that use VMI strategy may be able to reduce demand variability and thus their distribution and inventories carrying cost. Customer will benefit from product availability and high service level (Savelsbergh and Song 2007; Bard and Nananukul 2009) and thus can focus on their core business. However, the realization of the cost savings opportunities is very difficult, mostly with a large number of customers. The inventory routing problem that will be discussed later aims to solve this issue by determining a distribution strategy that minimizes long term distribution costs (Savelsbergh and Song 2007).

3.3 PRODUCTION, INVENTORY, DISTRIBUTION, ROUTING PROBLEM (PIDRP)

One of the main challenging problems faced by supply chain managers is the integration of production and distribution decisions when trying to optimize the whole supply chain. At the planning level, the objective is to coordinate production, inventory, and delivery to meet customer requirements in a way that minimizes the related costs (Bard and Nananukul 2010). In a vendor manage inventory relationship between the buyer and the supplier described previously, we defined a production, inventory, distribution, routing problem (PIDRP) as a coordination of the production schedules and inventory routing plans.

In general, the PIDRP seeks to coordinate the primary components of the whole supply chain (Bard and Nananukul 2009). It was first formulated by Lei et al. (2006) as a mixed integer program (MIP). They solved the problem with a two-phase solution approach that avoided the need to address lot-sizing and routing simultaneously. The PIDRP is a very complex problem that combines a capacitated lot-sizing problem with a capacitated multi-period vehicle routing problem. It has been proven to be beyond the capacity of exact methods (Bard and Nananukul 2010). In fact, the PIDRP is NP-hard in the strong sense.

When the main focus is on the routing planning, the PIDRP is most similar to the inventory routing problem (IRP) that will be discussed next and the periodic routing problem (PRP) (Bard and Nananukul 2010).

3.4 INVENTORY ROUTING PROBLEMS (IRP)

Given a production or distribution centre and a set of retailers or customer locations with their demand rates, the objective of the inventory routing problem (IRP), also called the One Warehouse Multi-Retailer Distribution Problem (Anily and Bramel 2004), is to determine a distribution plan that minimizes fleet operating and average total distribution and inventory holding costs without causing a stock-out at any of the sales-points during a given planning horizon. It is a challenging NP-hard problem that has been approached in different ways with respect to the inventory policy used at the sale points, the restrictions on service level, and the time horizon (Aghezzaf, Raa, and Van 2006).

Strategically, there are two type of inventory routing problem: The strategic IRP and the tactical IRP. The goal of the strategic problem is to estimate the long-term minimum size of the vehicle fleet required to serve the sales-points. The tactical problem focuses on routing an existing vehicle fleet to supply sales-points whose actual demands for replenishment can be estimated (Larson 1988).

In order to optimally get advantage of the vendor managed inventory relationship, Bard and Nananukul (2009) described three key decisions that must be made periodically:

1. Selection of customer to visit each period. It may be best to schedule a delivery for a stock replenishment rather than just to meet the periodic demand.
2. The quantity to deliver to each customer
3. The construction of routes.

3.4.1 DETERMINISTIC INVENTORY ROUTING PROBLEMS (DIRP)

Most of the IRPs have been solved by deterministic models, particularly for real size systems (Popović, Bjelić, and Radivojević 2011). Deterministic IRP consider parameters such as demand and travel time constant over the planning time horizon which is generally not the case for real problems. However, deterministic models are important and useful for many cases. Popović, Bjelić, and Radivojević (2011), studied the applicability of deterministic IRP solutions to stochastic problems with planning periods of different lengths using simulations. Their studied revealed that, solutions based on deterministic consumption can be applied in stochastic IRP by applying and balancing emergency deliveries and safety stocks measures.

3.4.2 STOCHASTIC INVENTORY ROUTING PROBLEMS (SIRP)

Few researches have employed stochastic programming in IRP. In order to find a solution for IRP with a single product, direct deliveries where customers are represented with probability distributions, Kleywegt et al. (2002) proposed Markov decision process and approximation methods. In their study on IRP for retailers that face stochastic demands over a long time period, Chen and Lin (2009) proposed a hedge-based stochastic inventory-routing system (HSIRS) with generalized autoregressive conditional Heteroskedasticity (GARCH) model for stochastic demand forecasting. The HSIRS was able to deal with a multi-product, multi-period replenishment policy with limited vehicle capacity and time window constraints.

3.5 GENERALIZED TRAVELING SALESMAN PROBLEM (GTSP)

The GTSP was introduced by Henry-Labordere, Saksena and Srivastava in the 1960s, in the context of computer record balancing and of visit sequencing through welfare agencies (Yang, 2008). The generalized travelling salesman problem (GTSP) is a more practical extension of the well-known travelling salesman problem (TSP). The TSP is one of the most famous problems in combinatorial optimization and has been used as a ground for many real life problems (Voudouris, 1999). It can be stated as a salesman spending his time to visit a given number of cities or nodes. In one tour, he visits each node once and end up where he started. The order of visit is done in a way that minimizes the total travel distance. For the GTSP, the nodes are a union of clusters with possible intersections between some of them. Each feasible solution, g-tour, is a closed path that includes at least one node from each cluster. It can be defined on a complete graph $G = (V, A)$, where $V = \{0, 1, ..., n\}$ is a vertex set and $A = \{(i, j) | i, j \quad V\}$ is an arc set. The set $V \setminus \{0\}$ is partitioned into a collection of vertex sets $\{C_1, C_2, ..., C_m\}$, where each vertex set C_k $1 \leq k \leq m$, is called a cluster. The objective is to find a g-tour with the minimum distance or cost. A special case of the GTSP is the E-GTSP where each cluster is visited exactly once. The GTSP is an NP-hard problem given that the TSP is its special case where the number of nodes within every cluster is equal to one.

The GTSP is a useful model for problems that involve decision of selection and sequence (Dimitrijević, 1997). For our problem, the nodes represent stands and clusters represent districts in the forest.

CHAPTER 4 DESCRIPTION OF THE WOOD CHIPS SUPPLY CHAIN

The use and the trade of fuel wood are increasing globally, which makes it a strategic resource for future energy. This is mostly due to the sustainability concern from some countries and partly from the pressure of the Kyoto Agreement to reduce emissions of greenhouse gases. The trade of emissions in Europe started in 2005 and there are now many systems with green certificates for electricity in countries such as Sweden. Wood fuel is produced both for domestic and industrial purposes. The major industrial use is for production of heat and electricity in stand-alone plants and in combined heat and power plants (CHP plants) (Bengt 2006).

Wood fuel can be divided into: firewood, sawdust and slabs, pellet and briquettes and wood chips. Firewood is round or chopped, split wood from cut-off root ends, delimbed stems and top and branches of softwood or hardwood. Sawdust and slabs are usually a by-product or residue from wood industries. Pellet and Briquettes are dry, comminuted wood consisting of shavings and sawdust compressed at high pressure.

Wood chips for which the work is focus on are obtained mostly from first and second thinning in spruce stands, harvesting over mature pine plantations and harvesting in climate and damaged stands. It result also from nurse trees and tops by clear-cutting in spruce stands.

Nurse trees are species planted at the same time of the primary tree species in order to protect them against frost and weeds. Clear-cutting is timber harvesting of the whole stand at the end of the rotation.

Wood chips are also produced from willow crops in short rotation forestry. But it has only been cultivated for a few years in Denmark. Presently, willow woods chips are only used to a limited extent at district heating plants. The willows can grow for at least 20 years without reduction in the plant yield which means that harvesting can take place 4 to 5 times before new planting is required. However, long-time storage of willow chips is difficult to handle. This is due to the high moisture content (approx. 55-58% of the total weight). In addition, the high moisture content makes the wood chips suitable only for plants equipped with a flue gas condensation unit (a particular method for heat recovery where a flue gas acts as a medium to warm the firing air). Long-term storage of un-chipped willow is better but expensive. For these reasons, willow wood chips are normally transported directly to the heating plant (Danish Energy Authority, 2002).

The increasing demand of energy wood is partly ensured by afforestation which includes the planting of new forests on agricultural land. This is usually done by increasing the number of plants compared to that of the normal stands, and by using nurse trees.

Wood chips are comminuted wood in lengths of 5-50mm in the fibre direction. The type of wood chips required depends on the class of the heating plant system.

The new standard subdivides the wood chips into five types according to their size grading:

Fine chips are appropriate for small domestic boilers where the chips are moved from the silo to the boiler with a screw conveyor. The screws are of a smaller dimension and very sensitive to large particles and slivers.

Coarse chips are suitable for larger boilers that are able to handle a coarser chip.

Extra coarse chips with a limited amount of fine material are appropriate for heating plants with grates where the chips are usually forced into the boiler.

Air spout chips are suitable for installations throwing the chips into the combustion chamber. These installations need a certain amount of "dust" and are sensitive to slivers.

Gassifier chip is an extra coarse type of chips with a very limited amount of "dust" and other fine particles. This type of chip is especially suitable for smaller gassifiers (Danish Energy Authority, 2002).

Wood chips have been mainly utilized for district heat production in small boiler plants in Denmark since 1980s. Nowadays, the consumption has gradually increased and the imported fuels are now competitive. In 2004, 90% of the wood chips burned were supplied domestically. Thirty-six percent of the wood chips were used in central power plants and 48% in decentralized co-generation of heat and power.

Logistics operations in Danish forests are expensive compare to other countries. In fact, forests are sparse, scattered and heat and power supply are highly decentralized. District heating plants are not essentially located close to forests, which leads to high transportation cost as well as environmental impact. There is a possibility for power plants to import wood fuel from neighbouring countries, but the longer travel distances may have an impact on the carbon balance. In addition, there are many associated negative effects such as road congestions and transport volumes. The better option is to minimize local fuels transport cost, which can be done by optimal utilization of vehicle payload and by choice of shortest travel paths. Transport cost alone constitutes 20% of the delivered wood chips costs.

The total forest area is about 486000ha of which 60% are coniferous and 40% deciduous. This complexity highlights the great importance of a carefully defined and implemented production and distribution system. On average, the yearly felling rate is about 4m^3/ha and the long term average annual biomass increase is 11m^3/ha. This is a clear improvement of the national forest plans that seek to double the forest coverage over the next 80 to 100 years.

Ninety five percent of all woods chips are produced from coniferous forest by chipping summer dried thinning or final fell residues. The moisture content decreased by drying is about 40 to 50 % (Möller and Nielsen 2007).

The moisture content is the most important physical parameter that determines fuel quality and is highly associated with species, the type and size of the tree and the drying time. The ash content (stone, soil and sand) in the whole trees depends on the type of the tree and on the quantity of unwanted branches and stem wood.

Low moisture contents increase the heating value of the wood chips and reduce the transportation cost. Generally, the properties of wood chips comprise a heat value of 10.4 GJ/wet tonne, a density of 0.25 wet tonnes/m^3 pr 1.4 wet tonnes/tonne of dry matter at moisture content of 40% (Möller and Nielsen 2007). The minimization of the losses of dry matter, substance that produces energy upon combustion, during storage is highly critical for the supplier of wood chips.

4.1 PRODUCTION

The actual production of wood chips is done after felling of whole trees and summer drying. Felling is usually done in the first three months of the year because the moisture content of the trees is lowest in that period. In fact, the wood chips produced should be as dry as possible. This also helps to reduce the risk of stump infection. The felled trees are then left in the area for drying in summer. Chipping is done by a chipper consisting of self-propelled basic machine with cabin, chipper and crane equipment mounted at the front part of the machine. It usually takes place directly at the roadside, at the storage terminal or sometime at the plant (see Figure 4.1).

4.1.1 CHIPPING AT ROADSIDE

The trees are felled and dried at the terrain and the un-chipped bio fuels are transported to the roadside for chipping. The chipper is a very expensive machine which has to be fully utilized. For that reason, a tractor with high-tipping trailer or a specialized forwarder is used to follow the chipper thereby enabling it to continue chipping while the forwarder carries the wood chips to the roadside storage.

4.1.2 CHIPPING AT THE TERMINAL

The felled and dried trees are stored at the supplier's storage of un-chipped bio fuels (districts and stands) and then transported to the terminal where chipping is done. After chipping, the wood chips are stored at the supplier's storage of wood chips or directly transported to the customer's outdoor of wood chips.

4.1.3 CHIPPING AT THE PLANT

We refer to the term plant here as district heating and power plants.

The whole dried trees are transported from the supplier's storage of un-chipped bio fuels to the storage of un-chipped wood fuel at the customer's facility (plant). The chipping is then done at that storage and transported to the customer's outdoor of wood chips.

Wood chips are required to be stored on hard surface and protected against rain. It is preferable to locate heating plants close to the source of fuel, which is not the case for most of the plants in Denmark. This is due to many reasons such as the heterogeneity distribution of the forests. The necessity to store the wood chips before shipping to the heating plants comes mainly from the following reasons:

- There is a high variety of wood chips during the year.
- Harvesting of wood chips is not possible for certain periods of time.
- More wood chips are produced than consumed in summer time.

As mentioned before, the forest in Denmark is scattered. For that reason, the tree-plant is subdivided into cluster (district) according to their geographical locations for the production (chipping). Each district contains a certain number of stands which in turn is composed with felled trees. The chipping rate and volume at stands are different because of the number and quality of felled trees at each stand.

Presently, the supplier is using only one chipper, which means that the chipper has to move from district to district and between stands within a given district to produce wood chips. There is a high cost associated with the relocation of the chipper between districts.

4.2 TRANSPORTATION

The transportation of wood chips is done by means of standard road trucks with trailer carrying two removable containers with a volume of $40m^3$ each. The containers are left at the roadside for loading. It is safely assumed that the volume of the containers corresponds to the allowed payload.

Wood chips are transported from roadsides or storage terminal to the outdoor or indoor storage of chips at the plants. In addition to the transportation of wood chips, un-chipped woods are transported from roadsides to the terminal or directly to the customer's storage of un-chipped bio fuel (at the plants).

The transportation costs mostly depend on the travel time, which itself depends on the distance and properties of the road. Hourly truck costs are used to calculate the transportation costs. The service time of trucks is assumed to be two times that of the travel time for the lead distance plus the time spent at the terminal or roadside (Möller and Nielsen 2007)

4.3 CONSUMPTION

The woods chips are supplied to district heating and CHP plants located in different geographical areas. The plants have different daily consumption rate and maximum storage capacity.

The technologies used at power plants specify various requirements with regard to the physical properties of the wood (size, size distribution, moisture content, ash content and pollutants (stones, soil and sand). District heating plants usually accept wood chips with a moisture content of 30 to 55%.

A measurement of the wood chips is required for sale price, which mainly depend on the quality and caloric value of the wood chips. The quality of the wood chips depends on the moisture content, the size distribution and on the impurities (soil, stone etc.). The handling and burning properties are usually the most factors considered when assessing the quality of the wood chips. In fact, a poor wood chip quality often leads to difficulty in handling. In addition, the wood chip quality also has an influence on the combustion efficiency and on the content of harmful substances in smoke gas and ash.

The caloric value is the number of heat units obtained either per weight or volume unit by the complete combustion of a unit mass of fuel.

The net caloric value is the most caloric value used in Denmark and forms the basis of the sale and purchase of wood chips. Net calorific value can be defined as the units of heat produced by the complete combustion of a well-defined amount of wood chip with the moisture content in the wood and the vapour that is formed during combustion (approx. 0.5 kg water per kg dry matter) being in a gaseous state. In fact, the recovery of heat by condensing the vapour in the flue gas is not included. The effect of the moisture content on the caloric value can be calculated by the following formula: $H_{n,v} = H_n \left(\dfrac{100 - F}{100} \right) - \dfrac{2.442F}{100}$

where:

> $H_{n,v}$ is the net calorific value of wet wood (GJ per tonne total weight)
> H_n is the net calorific value of dry wood (GJ per tonne total weight)
> F is the moisture content in percentage of total weight
> 2.442 is the latent heat of evaporation of water at 25°C (GJ per tonne)

The payment of wood chips by most Danish chip-fired heating and CHP plants is based on the energy content determined as the net calorific value per tonne of total weight.

The net caloric value presented in the formula above has different conversion according to the type of the wood chip:

For wood chips from Scandinavian region consisting mainly of pine, spruce and birch wood, $H_{n,v} = 19.2 - 0.2164 \times F$ (GJ per tonne total weight) where F is the moisture content of the wood chips in percentage of the total weight of the wood chips.

For mixed wood chips from various origin consisting largely of hardwood of unknown mixture, $H_{n,v} = 19.0 - 0.2144 \times F$ (GJ per tonne total weight) where F is the moisture content of the wood chips in percentage of the total weight of the wood chips.

Below is the calculation example for softwood forest chips:

- Moisture content in wood chips: 55% of total weight
- Weight: 15 tonnes
- Energy price (1998): DKK 35.00/GJ
- Wood chip calorific value Hn,v: 19.2 GJ/tonne - (0.2164 × 55) = 7.30 GJ/tonne
- Wood chip energy content: 15 tonnes ×7.30 GJ/tonne = 109.50 GJ
- Wood chip price: DKK 35.00/GJ ×109.50 GJ = DKK 3,832.50 (Danish Energy Authority, 2002).

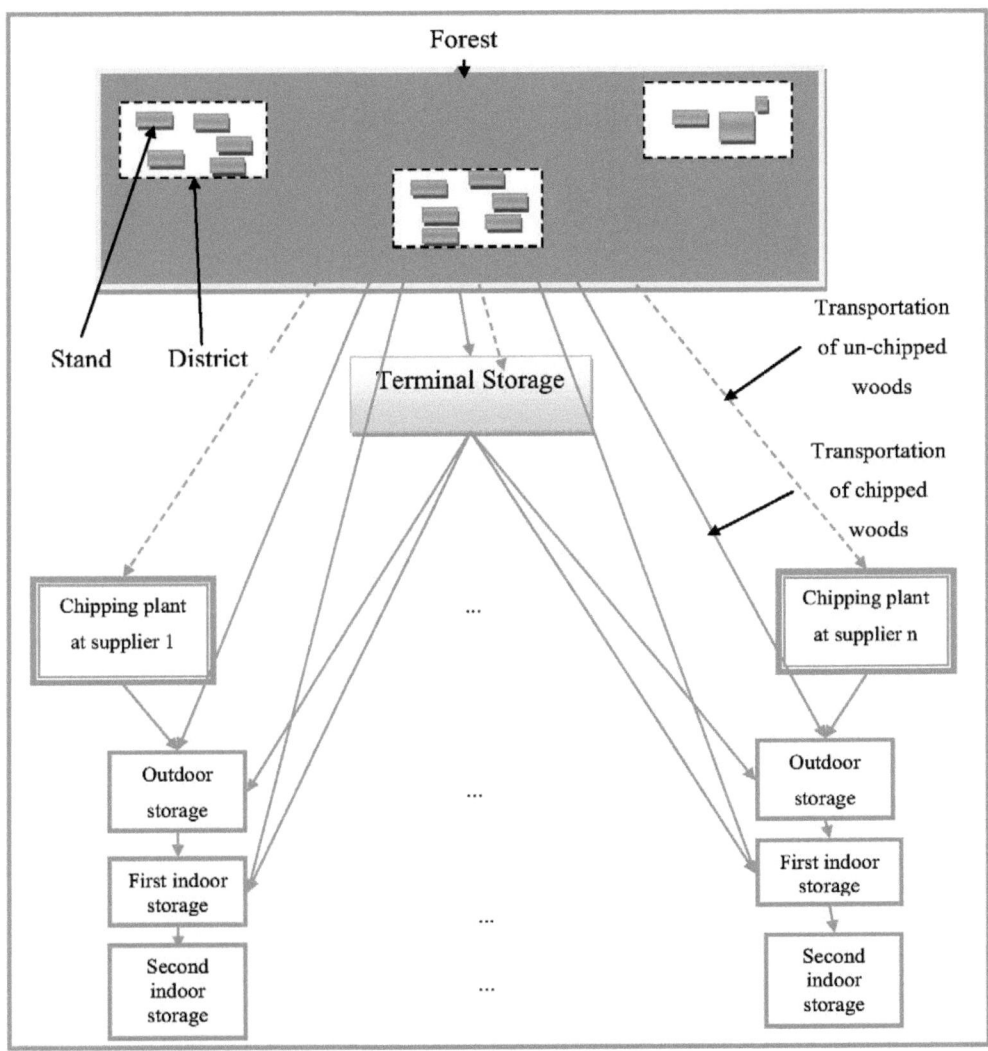

FIGURE 4.1: WOOD CHIPS

4.4 A MIP MODEL FOR THE PIDRP

A mixed integer programming (MIP) model developed to describe the problem is based on the vehicle routing with time windows model from J-F, Laporte, and Mercier (2001) and from the inventory production model developed by Bard and

21

Nananukul (2010). The transportation of un-chipped wood is not explicitly included. In addition, we have considered the relocation cost of the chipping machine(s) as setup cost in the production.

4.4.1 PARAMETERS AND VARIABLES

Indices and sets

i, j	Indices for customers (power and heating plants)
k	Index for districts (the production in the forest in subdivided in distircts)
t	Index for periods
N	Set of customers
K	Set of clusters
T	Set of periods in the planning horizon; $T_0 = T \cup \{0\}$ and $|T| = \tau$

Parameters

d_{it}	Demand of customer i in period t
D_{it}^{max}	Upper bound on the maximum amount to be delivered to customer i in period t
D_t^{max}	Upper bound on the maximum amount that can be loaded on a vehicle in period t
γ	Number of available vehicles
Q	Capacity of each vehicle
I_{max}^p	Maximum inventory that can be held at the terminal (storage)
$I_{max,i}^c$	Maximum inventory that can be held at the heating plant i
C	Periodical production capacity at the forest
c_{ij}	Travelling cost from customer i to customer j

22

f_t	Total cost of relocating the production between stands (that belong to different districts) used in period t (Cost of Hamiltonian path)
h^p	Unit holding cost at the terminal (storage)

Decision variables

x_{ijt}	1 if customer i immediately precedes customer j on a delivery route in period t; 0 otherwise
y_{it}	Load on a vehicle immediately before making a delivery to customer i in period t
p_t	Total production quantity in period t
z_t	1 if there is production in period t; 0 otherwise
I_t^p	Inventory at the terminal facility at the end of period t
I_{it}^c	Inventory at customer i at the end of period t
w_{it}	Amount delivered to customer i in period t

4.4.2 THE MODEL

$$\text{minimize} \sum_{t \in T} \sum_{i \in N \cup K} \sum_{j \in N \cup K} c_{ij} x_{ijt} + \sum_{t \in T} f_t z_t + \sum_{t \in T \setminus \{\tau\}} h^p I_t^p$$

Subject to :

$$I_t^p = I_{t-1}^p \quad p_t + \sum_{i \in N} w_{it}, \quad t \, \forall T \in \tag{1}$$

$$I_{it}^c = I_{i,t-1}^c \quad w_{it} + d_{it}, \quad i \, \forall N \in t \, \forall T \tag{2}$$

$$\sum_{i \in N} w_{it} \leq I_{t-1}^p, \quad \forall t \in T \tag{3}$$

$$p_t \leq C z_t, \quad \forall t \in T \cup \{0\} \setminus \{\tau\} \tag{4}$$

23

$$p_0 \geq \sum_{i \in N} \left(d_{i1} - I_{i0}^c \right) - I_0^p \tag{5}$$

$$\sum_{\substack{j \in N \cup K \\ j \neq i}} x_{ijt} \leq 1, \ \forall i \in N, t \in T \tag{6}$$

$$\sum_{\substack{i \in N \cup K \\ i \neq j}} x_{ijt} = \sum_{\substack{i \in N \cup K \\ i \neq j}} x_{jit}, \quad i \quad N, t \forall \ \mathbb{E} \tag{7}$$

$$\sum_{j \in N} x_{kjt} \leq \gamma, t \in T, k \in K \tag{8}$$

$$y_{jt} \leq y_{it} - w_{it} + D_t^{\max} \left(1 - x_{ijt} \right), \ i \in N, j \in N \cup K, t \in T \tag{9}$$

$$w_{it} \leq D_{it}^{\max} \sum_{j \in N \cup K} x_{ijt}, \ i \in N, t \in T \tag{10}$$

$$0 \leq I_t^p \leq I_{\max}^p, 0 \leq I_{it}^c \leq I_{\max,i}^c, \forall i \in N, t \in T \setminus \{\tau\}; I_\tau^p = I_{i\tau}^c \quad 0;$$
$$x_{ijt} \in \{0,1\}, 0 \leq y_{it} \leq Q, w_{it} \in \square_+^{nx\tau}, \forall i \neq j \in N \cup K, t \in T; \tag{11}$$

$$z_t \in \{0,1\}, p_t \in \square_+^\tau, t \in T \cup \{0\} \setminus \{\tau\}, p_\tau = 0$$
$$D_{it}^{\max} = \min \left\{ Q, \sum_{l=t}^\tau d_{il} \right\} and \ \ D_t^{\max} = \min \left\{ Q, \sum_{i \in N} \sum_{l=t}^\tau d_{il} \right\} \tag{12}$$

4.4.3 MODEL DESCRIPTION

The objective function minimizes the total transportation cost, the relocation cost and the total inventory cost at the terminal. Constraints (1) and (2) are inventory balance constraints at the terminal and for all the customers. The planning starts with initial inventory at each heating plant. Constraint (3) insures that the amount of wood chips available for delivery to heating plant in a given period has to be equal to the inventory at the terminal in the previous period. Constraint (4) gives an upper bound on the amount to be produced in each period. Constraint (5) insures the initial inventory is sufficient to meet the demand in the first period. The routing for each period and vehicle is controlled by constraints (6)-(9).

24

The maximal amount to be delivered to the heating plant i when visited is controlled in constraint (10).

The number of vehicle that leaves the terminal or forest is limited in constraint (10). Constraints (11) and (12) give a definition and bound on the variables to be used.

CHAPTER 5   SOLUTION METHODS

5.1 EXACTS METHODS

Exacts solutions methods find the optimal solution of a given problem. The IRP, consequently the PDIRP, is very complex to be solved optimally. Exacts methods, such as branch and bound, can only be used for small instances (Amir and Azad 2010). The research conducted by Bard and Nananukul (2009) showed that PIDRP instances with up to eight time periods and 50 customers can be solved within 1 h.

Few researches on large instances have been conducted but with limits on the solution procedure. Christiansen (1999) and Al-khayyal and Hwang (2007) proposed exact methods with a limit on the solution time and reported the best found solution. Bell et al. (1983) used branch and bound with very limited branching.

5.2 HEURISTICS METHODS

Heuristics methods are sophisticated algorithms that find good or close to optimal solutions within a reasonable time. They are manly used to solve IRP as well as PIDRP. Amir and Azad (2010) established a heuristic method based on a hybridization of Tabu Search and Simulated Annealing. Their results showed that the proposed heuristic is considerably efficient and effective for a broad range of problem sizes. Aghezzaf, Raa, and Van (2006) suggested a column generation based approximation method to solve the nonlinear mixed integer formulation of their IRP. Various modern heuristics   such   as tabu search (Liu and Chen 2011), variable neighbourhood tabu search (Liu and Lee 2011) and A hybrid genetic algorithm (Moin, Salhi, and Aziz 2011) as well as simulation analysis (Popović, Bjelić, and Radivojević 2011), have been recently used by academicians in the literature to solve the IRP and its variants.

Bard and Nananukul (2010) developed a hybrid methodology that combines exact and heuristic (column generation) procedures with a branch-and-price for an integrated production and inventory routing problem.

CHAPTER 6 PROBLEM DECOMPOSITION

6.1 SCOPE AND SPECIFICS OBJECTIVES

In order to better approach the problem, we have subdivided the solution methods and processes into different parts. First of all, we consider the production plan which itself is divided into two parts: the yearly and the weekly production plan. Then, we consider the transportation of wood chips and un-chipped woods from districts to the terminal storage and to the customer's storage of un-chipped woods. Finally the daily or weekly inventory of wood chips replenishment to the customers.

Due to time constraint, the work is mainly focused on the yearly and weekly production plan. In addition, because of the time restriction, we do not consider the influence of the production sequence on the transportation cost of wood chips to customers.

As mentioned in chapter 4, the supplier only uses one chipper to produce wood chips. The chipper moves from districts to districts and between stands within a given district for production. In addition, every district has its own number of stands and consequently its own available production volume. Each stands belong to a specific district and has its own available production volume and production rate (type of felled trees).

For a given year, the production plan is done to satisfy the total customer's weekly demand. The specifics objectives for the production plan are the followings:

1. To determine the optimal number of woods chips to be produced and to be kept as inventory each week.
2. To optimally determine which districts and which stands to visit for a production each week and day.
3. To determine the optimal production sequence between districts (inter district production sequence).

4. To determine a close to optimal production sequence between stands belonging to the same district (intra-district production sequence) for each weekly production plan.

5. To optimally determine at which time (hour in a day) the production should be run at a given district and stand in the planning horizon.

We mainly use exact method to address the above objectives, except the fourth objective which is met by means of a heuristic method.

Clearly, we have developed a mathematical model that will address the objectives 1 to 3 and 5, then we use a tabu search with 2-opt best improvement to meet the fourth objective (see Figure 6.1).

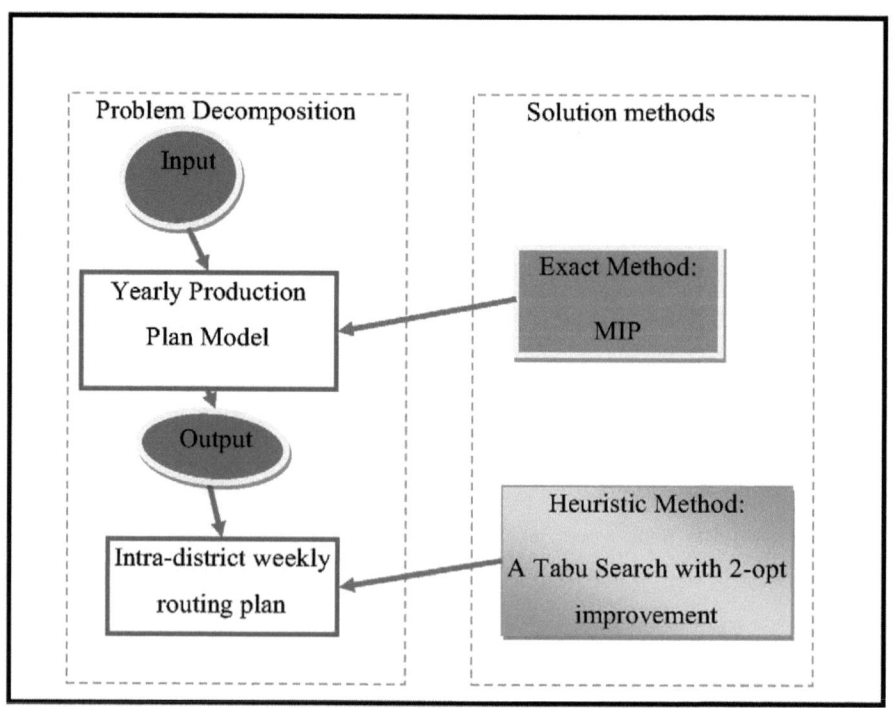

FIGURE 6. 1: PROBLEM DECOMPOSITION AND SOLUTION METHODS

6.2 A MIP MODEL FOR THE YEARLY PRODUCTION PLAN

The model we have developed is based on the description of the problem given previously. We define parameters which are input data to the model and variables to contain the output of the model. The main objective is to minimize the holding and the production routing cost. The plan is for one year and one week represent one period.

6.2.1 PARAMETERS AND VARIABLES

Sets

K	Set of districts in the forest
T	Set of periods (weeks) in the planning horizon (one year); $T_0=T \cup \{0\}$ and $|T|=\tau$
S	Set of stands

Parameters

d_t	Total customer's demand in week t (m^3)
r_{ij}	Production rate at stand i of district j (m^3/hour)
c_{ij}	Transportation cost between district i and j (NOK)
e_{ij}	Travel time between district i and j (hour)
L	Loading time of the chipper (hour)
U	Unloading time of the chipper (hour)
f	Fixed average travel time between stands (hour)
v_i	Available production volume at district i (m^3)
a_{ij}	Available production volume at stand i of district j (m^3)
ρ	Initial inventory
h	Unit holding cost at the storage
G	Number of working hour per week

M	A very large number

Decision variables

Y_{it}	1 if production is run at district i in week t,
	0 otherwise
Z_{ijt}	1 if production is run in stand i of district j in week t,
	0 otherwise
V_{ijt}	1 if the chipper move from district i to district j in week t
	0 otherwise
R_{ijt}	1 if production ends at district i in week t and start at district j
	in week t+1,
	0 otherwise
W_{it}	Time in which the production start at district i in week t
I_t	Inventory at the storage facility at the end of period t

6.2.2 THE MODEL

$$\min \sum_{t\in T} hI_t + \sum_{t\in T}\sum_{i\in K}\sum_{\substack{j\in K \\ j\neq i}} c_{ij}V_{ijt} + \sum_{t\in T\setminus\{\tau\}}\sum_{i\in K}\sum_{j\in K} c_{ij}R_{ijt}$$

St

$$I_t = I_{t-1} \sum_{j\in K}\left(\sum_{i\in S} a_{ij}Z_{ijt}\right) - d_t \qquad \forall t \in T \tag{1}$$

$$\sum_{t\in T} Z_{ijt} \leq 1 \qquad \forall i \in S \ \forall j \in K \tag{2}$$

$$W_{jt} + \left(\sum_{i\in S}\left(\frac{a_{ij}}{r_{ij}} + f\right)Z_{ijt} - fY_{jt}\right) + \left(e_{jk} + l + u\right) - W_{kt} \leq M\left(1 - V_{jkt}\right) \tag{3}$$
$$\forall j,k \in K : j \neq k, \forall t \in T$$

30

$$W_{jt} + \left(\sum_{i \in S} \left(\frac{a_{ij}}{r_{ij}} + f \right) Z_{ijt} - fY_{jt} \right) + \sum_{k \in K} \left(e_{jk} + l + u \right) R_{jkt} \le g \tag{4}$$

$$\forall j \in K, \forall t \in T$$

$$W_{jt+1} \le \left(1 - R_{ijt} \right) M \quad \forall i, j \in K \ \forall t \in T \setminus \{\tau\} \tag{5}$$

$$\sum_{i \in S} a_{ij} Z_{ijt} \le v_j Y_{jt} \quad \forall j \in K \ \forall t \in T \tag{6}$$

$$\sum_{i \in K} \sum_{j \in K} R_{ijt} = 1 \quad t \quad T \forall \{ \tau \} \tag{7}$$

$$\left(\sum_{i \in K: i \neq j} V_{ijt} + \sum_{i \in K} R_{ijt-1} \right) + \left(\sum_{i \in K: i \neq j} V_{jit} + \sum_{i \in K} R_{jit} \right) \ge 2Y_{jt} \quad \forall j \in K \ \forall t \in T \tag{8}$$

$$2V_{ijt} \le Y_{it} + Y_{jt} \quad \forall i, j \in K : i \neq j, \forall t \in T \tag{9}$$

$$2R_{ijt} \le Y_{it} + Y_{jt+1} \quad \forall i, j \in K \ \forall t \in T \setminus \{\tau\} \tag{10}$$

$$\sum_{i \in K: i \neq j} V_{ijt} + \sum_{i \in K} R_{ijt-1} = \sum_{i \in K: i \neq j} V_{jit} \quad \sum_{i \in K} R_{jit} + \quad j \quad K \ \forall t \in T \tag{11}$$

$$I_t \ge 0 \ \forall t \in T_0;$$
$$Y_{it} \in \{0,1\}, W_{it} \ge 0 \ \forall t \in T \ \forall i \in K$$
$$R_{ijt} \in \{0,1\} \ \forall t \in T_0 \ \forall i, j \in K$$
$$Z_{ijt} \in \{0,1\} \ \forall i \in S \ \forall j \in K \ \forall t \in T \tag{12}$$
$$V_{ijt} \in \{0,1\} \forall t \in T \ \forall i, j \in K : i \neq j; \ g{=}40$$
$$I_0 = \rho$$

6.2.3 Model Description

The objective function minimizes the total inventory holding cost and transportation cost of the chipper between districts in the forest.

Constraints (1): Inventory balance constraints

Inventory at the end of any period should be equal to the sum of the production volume in that period and the inventory in the previous period minus the demand of that period.

Constraints (2): Production constraints

Any stand should not be visited more than once for chipping and all the felled trees in a stand should be chipped if visited. The visited stands in a given period may belong to different district in the forest. Thus, more than one district may be visited in a period. In addition, the total volume of wood chips produced at any district in the planning period will not exceed its available production volume.

Constrains (3) to (5): Production time-sequence constraints

The production time at any district is equal to the sum of the production time at the visited stands plus the time used to move between stands. For a given period, the time at which the production start in a visited stand should be equal to the end of production time of the previously visited stand plus the travel time, including loading and unloading time, between the two stands. Constrains (4) insure there is enough time to move the chipper from the last visited district at the end of a given period, to the first visited district at the beginning of the next period. This time will be equal to zero if the chipping starts at the same district in the next period. The time is reset to zero at the beginning of each week in constraints (5).

Constraints (6) to (11): Production-routing constraints

The route is built between visited districts in the same week as well as those visited in the different weeks and only one chipper is used for the chipping. For a given visited district in a period, there should be only one arc coming in and one arc going out.

Constraints (12) give a definition and bound on the variables to be used as well as the initial inventory value.

6.3 A TABU SEARCH BASED METAHEURISTIC FOR INTRA DISTRICTS ROUTING

Tabu search (TS) is a local search based Meta heuristic method first introduced by Glover in 1986. It has successfully been used in the literature to solve various difficult combinatorial problems. Tabu search helps to move from local optimum that the pure local search is always trapped into. It starts with an initial solution s_0 and explores the solution space by moving, at each iteration, from current solution s to the next current solution s' which is the best solution of a subset $N(s)$ (neighborhood of s). The cost $c(s')$ of a solution s' might be larger than $c(s)$ of a solution s.

For that reason, it will be necessary to implement an ant-cycling mechanism. Particularly, if *TabuList* denotes a list of all forbidden solutions attributes, then any solution s' with attribute in that list will not be allowed as the next current solution for a certain number of time TT (Tabu Tenure) unless $c(s') \leq c(s)$ which is called *aspiration criteria*. Some other strategies such as *diversification* helps to search unexplored regions and *intensification* to explore the area of the search space that seem more promising.

These strategies use long term memories such as *frequency based memory*. The process stop after a predefined termination criterion is satisfied.

For each visited district in a given period (week), we run the developed 2-opt tabu search with best improvement algorithm to find a production sequence between selected stands of that district with reasonable total distance, if not minimum.

6.3.1 SOLUTION REPRESENTATION

We represent a feasible solution as a sequence of stands where each stand appear only once and in the visiting order. The first stand to visit is the one with smallest distance or road access with a previously visited district in the weekly plan presented in section 6.2 and the last visited node is the one with smallest distance or road access with the next district to visit from the weekly plan. In fact, the weekly plan gives the order for which the chosen district should be visited and a tabu search gives the order in which the selected stands should be visited. It is then obvious that a tabu search solution for a given district is a Hamiltonian path where the first and last visited stands are known. Figure 6.2 illustrates a solution sequence with 7 stands.

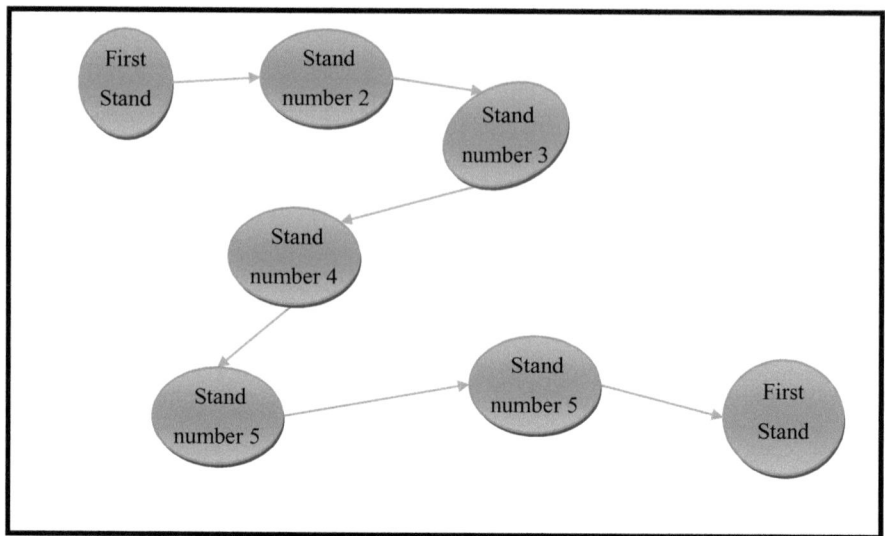

FIGURE 6.2: SOLUTION REPRESENTATION

6.3.2 MOVES AND NEIGHBORHOOD

As mentioned before, we use a 2-opt method to find the neighbor of a given solution. A 2-opt move replace two non-adjacent edges $e_{i,j}$ and $e_{m,n}$ by the two other edges $e_{i,m}$ and $e_{j,n}$ (Figure 6.3).

In order to maintain the route sequence, the sub path $(j,...,m)$ is reversed to $m,...,j$. The solution cost of the move can be expressed as:

$\Delta_{i,m} = dist(i,m) \quad dist(j,n) \quad (dist(i,j) \quad dist(m,n))$ where $dist(i,j)$ is the distance between stand i and j.

The neighbourhood $N(s)$ of a solution s is the set of all the solutions obtained by applying 2-opt moves in s. The cardinal of the set $N(s)$ is equal to

$$\binom{\binom{n}{2}}{2} - (n-3)$$

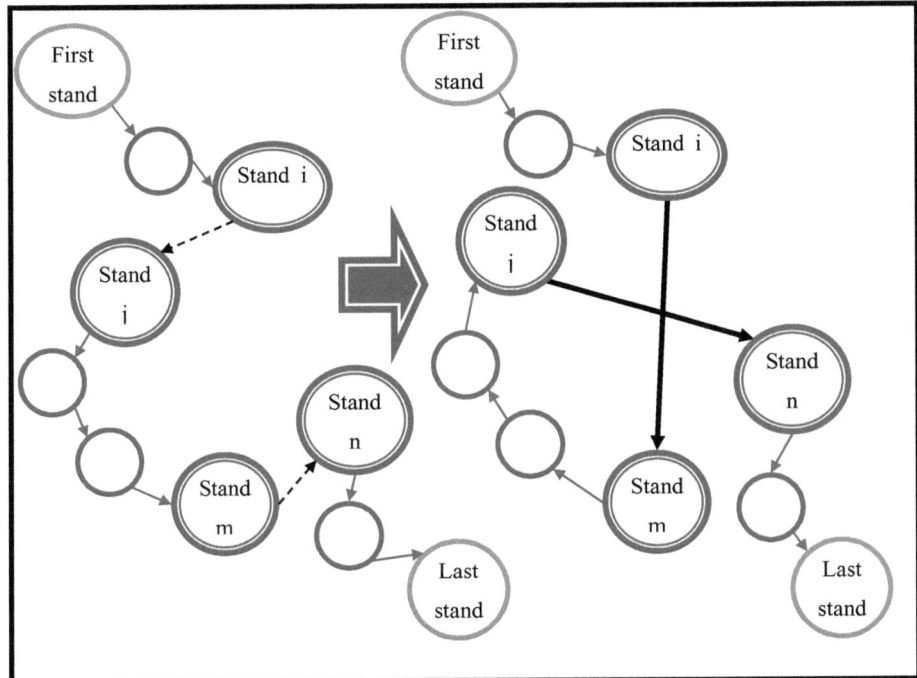

FIGURE 6.3: 2-OPT MOVE

6.3.3 Initial solution

A greedy approach is used to find an initial solution. The algorithm starts with the first stand and finds the nearest stand which will become the current stand. The nearest unvisited stand from the current stand is taken as the next current stand until all the required stands are visited.

6.3.4 Tabu list and aspiration criteria

In order to prevent the algorithm from cycling, the entering edges (distances) in a current solution are marked as tabu for a short number of iteration.

Aspiration is done by accepting a tabu solution with an objective value better than the best previously found solution.

6.3.5 PROPOSED ALGORITHM

The proposed TS algorithm is outlined in table 6.1. The notations used in the algorithm are explained below:

R^0	Initial route
R^k	Route at iteration k
$NB(R^k)$	The neighborhood of the current route
R^{best}	The best route found during the whole algorithm
$R^{best-tabu}$	The best route in the current neighborhood that can be reached by a tabu move
$R^{best-nontabu}$	The best route in the current neighborhood that can be reached by a non tabu move
$d(R)$	The total distance of the route R
d^{best}	The distance of the best route found during the whole search
TL	Tabu list
TT	Tabu tenure
α	Number of iteration

Strategies such as diversification and intensification with frequency based memories were not used because the number of stands to visit a week at a given district is not very large. In addition, the most important distance minimization concern was optimally solved in section 6.3.

The algorithm uses full enumeration for districts with a number of stands lower or equal to eight. That is, the optimal solution is guaranteed for those cases.

Initialization
Generate an initial random route.
set $R^{best} = R^0$, $d^{best} = d(R^0)$.

$TL = empty$.

Iteration number $k = 0$.

Step 1
Create $NB(R^k)$ *and invoke the evaluation function for each neighbor of the current solution to determine the potential new better route.*

Step 2
1. If $d(R^{best}) \leq \min\{d(R^{best-tabu}), d(R^{best-nontabu})\}$

$\quad$ set $R^{k+1} = R^{best-nontabu}$,

2. If $d(R^{best-nontabu}) \leq \min\{d(R^{best-tabu}), d(R^{best})\}$

$\quad$ set $R^{k+1} = R^{best-nontabu}$, $R^{best} = R^{k+1}$, $d^{best} = d(R^{k+1})$,

3. Aspiration: If $d(R^{best-tabu}) \leq \min\{d(R^{best-nontabu}), d(R^{best})\}$

$\quad$ set $R^{k+1} = R^{best-tabu}$, $R^{best} = R^{k+1}$, $d^{best} = d(R^{k+1})$.

Update the TL for TT number of iterations with the edges that produced R^{k+1}

Step 3: *Set* $k = k$ ⊣ *and go to* **step 1** *until the termination criteria is satisfied*

TABLE 6. 1: PROPOSED TABU SEARCH ALGORITHM FOR THE INTRA DISTRICTS ROUTING

6.4 IMPLEMENTATION

The developed mathematical model is implemented in AMPL with IBM ILOG CPLEX Optimization Studio System 9.0 as a solver and a Tabu Search algorithm in C++ using Microsoft Visual Studio 2010.

The results from AMPL needed as input to a Tabu Search algorithm are saved in well-formatted files. To put it bluntly, the selected stands for the production of woods chips for every week are saved as input files for the intra-routing algorithm.

The first and the last stand to visit at a given district are determined in advance. In fact, the first stand to visit in a given district is the one with a smallest distance from the previously visited district and the last stand to visit is the one with the smallest distance to the next district to visit in the inter districts production route. A district is consider as a cycle with a center point and the distance from any stands to a given

37

district is equal to the distance from that stand to the center point of the related district.

We also make use of CONCORD software online and our own implemented TSP with fixed starting and finishing nodes to compare the result with a Tabu Search algorithm. In addition, SPSS and EXCEL are used to analyze the raw and output data respectively.

6.5 DATA COLLECTION

The input data needed to run the model is presented in table 6.2. The last column of the table shows the data provided by Forest and Landscape Denmark. In fact, we had a meeting with an employee from FLD at Molde University College where some critical information were given. An excel sheet containing most of the production information from 2002 to 2011 was also made available with other documents.

However, the information regarding the distances, geographical location, average speed of the chipper to move between stands and districts was not provided as well as the demand and the unit holding cost. It was not possible to have that information before the due date.

Tables containing the data input are presented in the appendix section. We have mainly used the production information from May 2010 to May 2011 to run the model.

We have run ANOVA test statistic on SPSS to investigate the strength of the association between *tree species (quality)* and the *production rate* at each stands and districts (see Appendix). The result shows a significant relationship. Clearly, the production rate is different for each stands, depends on the quality and type of the felled tree.

Estimations of the information not provided in due time are presented in the appendix section as well.

Input data	Unit measure	Availability
Total weekly customer's demands	m^3	Not provided
Production rate at each stand	m^3/h	Provided
Transportation cost and distance between districts and between stands	NOK and Km	Not provided
Average speed of the chipper to move between stands and districts	Km/hour	Not provided
Loading and unloading time of the chipper	Hour	Not provided
Average production volume at each stand	m^3	Provided
Unit holding costs	NOK	Not provided

TABLE 6. 2: REQUIRED INPUT DATA

CHAPTER 7 COMPUTATIONAL EXPERIMENTS

7.1 TEST CASES

Apart from the real-world instances, we have randomly generated small and large instances to run the model and a Tabu search algorithm. Instances were generated as follow:

Demands

The weekly demands were created between a defined minimum, mean and standard deviation. The demands for winter were considerably higher compare to summer as to follow the real demand pattern.

Production rate and volume at stands

The production volumes were randomly generated within defined minimum and maximum values. In addition, the production volumes were defined by taking the total yearly demand into consideration. We have defined three types of woods species (qualities) which have different production rates. Each stand was associated to wood specie with equal probability. The number of stands and districts were fixed and adjusted based on the instance size needed.

Stands and districts locations

Stands were randomly associated with districts and distances between districts are larger compare to distances between sands within a given district.

7.1.1 TEST CASES FOR SMALL SIZE INSTANCES

To run the MIP model with small size instances, we have generated the input data presented in table 7.1 and table 7.2. A total of 30 (1 to 30) stands were randomly associated to 5 districts (A, B, C, D and E). The three types of tree species (quality) have a production rate of 50, 100 and 200 m^3 (see table 7. 1). The distance (in km) matrix between stands together with the other important input parameters are presented in table 7.2. The unit holding cost is adjusted based on the specific small

40

instance case. In fact, we are more concerned about the behavior of the model with different unit inventory carrying costs. We have decided to take 6 time periods (weeks) where week 3 and 4 represent the peak periods (winter time). The related weekly demands are presented in the results section. We have decided to use small volume and distance values to make the cases less difficult to read and make comparisons.

Distance Matrix					
	A	B	C	D	E
A	-	0.57	0.71	1.31	0.57
B		-	1.28	1.68	0.87
C			-	1.22	0.88
D				-	0.81
E					-
Parameters values					
Speed	30km/hour				
Travel cost	300NOK/hour				
Weekly working capacity	40 hours				
Looading time	1 hour				
Unloading time	2 hours				
Average travel time between stands	1 hour				
Unit holding cost	200 NOK/unit				

TABLE 7. 1: DISTANCE MATRIX BETWEEN DISTRICTS AND PARAMETERS VALUES

Stand number	Related district	Production rate	Production volume (m3)
1	C	200	50.78
2	B	200	28.24
3	D	50	37.58
4	D	200	41.63
5	C	200	38.54
6	B	200	37.27
7	A	200	44.5
8	D	100	39.19
9	D	50	36.83
10	B	200	55.52
11	D	200	28.83
12	C	100	42.28
13	B	100	32.03
14	D	50	35.67
15	D	100	48.9
16	C	200	27.74
17	D	100	38.7
18	A	50	10.55
19	A	200	38.37
20	B	50	18.81
21	A	200	31.91
22	B	100	45.88
23	C	100	46.29
24	D	50	34.66
25	B	100	40.33
26	D	200	37.03
27	D	100	36.72
28	D	100	30.68
29	D	50	31.39
30	E	200	55.02

TABLE 7. 2: PRODUCTION RATE AND VOLUME AT EACH STAND

7.1.2 TEST CASE FOR LARGE SIZE INSTANCES

The test case for a large size instance matches with real world instances except for the few type of tree species which are set to three. We have used 21 districts, 1000 stands and 52 weeks.

7.1.3 TEST CASE FOR REAL-WORLD INSTANCES

The real world instance chosen is that of May 2010 to May 2011 with 21 districts, 1734 stands and 34 different tree species.

7.2 RESULTS AND ANALYSIS

7.1.1 TEST RESULTS AND ANALYSIS FOR SMALL SIZE INSTANCES

Three instances with different inventory costs were tested to evaluate the influence of the unit carrying cost on the solution output. Actually, the holding cost is the main concern of the supplier of wood chips in Denmark. However, by adjusting the unit holding cost, the travel cost of the chipper is implicitly adjusted as well.

High inventory holding costs

In this case, the concern in more on the holding cost minimization. Consequently, the unit inventory carrying cost is set to 200NOK/unit. In addition, the average inventory level with the current supplier's policy is equal to 20% of the production volume. This is why we have included the corresponding inventory level in order to make comparisons with our results (see table 7.3).

The plan in table 7.3 shows the stands (with their corresponding districts) to visit for the production each week. For example, the stand number 7 located at district number A will be used for the production in the first week. The stand number 29 in district number D will be used in the fourth week which represents the last part of the winter time (week 3 and 4 represent winter time).

The volumes of wood chips to be produced each week as well as the inventory level at the end of each week are presented in table 7.3. A high inventory cost gives an output plan with high number of stands in different district to visit for the production. Four districts out of five are visited for the production.

Week	1	2	3	4	5	6
Demand (m³)	44	98	271	284	91	42
Stand used	7A	19A, 20B, 25B	18A, 2B, 22B, 12C, 16C, 15D, 27D, 28D	3D, 8D, 9D, 11D, 17D, 24D, 26D, 29D	10B, 14D	4D
Production volume (m³)	44.5	97.51	270.99	284.21	91.19	41.63
Inventory level (m³)	0.5	0.01	0	0.21	0.4	0.03
20% inventory (m³)	27.66667	27.66667	27.66667	27.66667	27.66667	27.66667

TABLE 7. 3: RESULTS FOR HIGH INVENTORY CASE INSTANCE

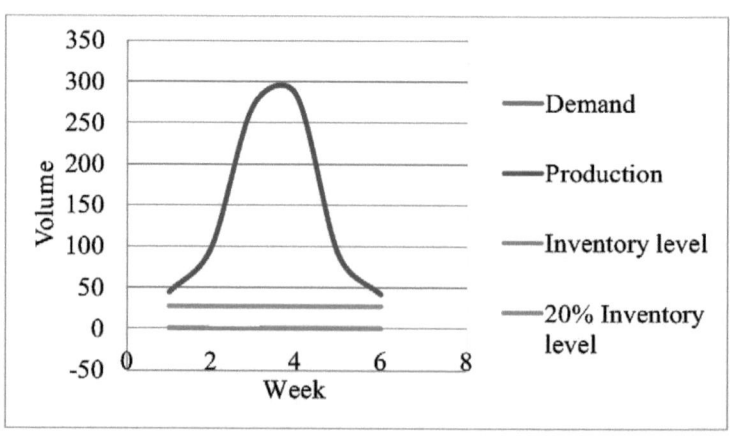

FIGURE 7. 1: DEMAND AND INVENTORY CURVES

As shown in figure 7.1, the production curve is almost equal to that of the demand curve. That is, we try to produce the volume that will satisfy the demand of the week with fewer inventories, if not zero.

In addition, the inventory level line is almost at zero which is by far below that of 20% inventory level used in a supplier's current policy (see figure 7.1). The maximum inventory level in this case is 1.1 % of the production volume.

The peak season is well illustrated in figure 7.1. The demand is low in period 1 and 6, fairly high in period 3 and 5 and very high in period 3 and 4.

The schedule for the whole production planning in the time horizon is presented in table 7.4. The production starts in week 1 on Monday 8.00 o'clock at district number A and ends in week 6 at district number D.

There is no production at any stands in district E for this planning. That is, some districts or stands may never be visited for the production. As highlighted in the further research section, a constraint can be easily added in the model to visits some specific districts or stands which have not been visited for a certain fixed time.

Schedle			Week			
District	1	2	3	4	5	6
A	Monday 8.00am	Monday 8.00am	Monday 2.10pm			
B		Monday 1pm	Monday 8.00am		Tuesday 8.00am	
C			Tuesday 8.00am			
D			Wednesday 8.00am	Monday 8.00am	Monday 8.00am	Monday 8.00am
E						

TABLE 7. 4: PRODUCTION SCHEDULE

The blue sequence in figure 7.2 shows the route followed by the chipper to run the production between districts. The chipper start the production at district number A then moves to district number B then goes back to district A, C , D then moves to district B and complete the production at district D. The route in blue color is the smallest Hamiltonian cycle. It can be seen that the chipper will follow that cycle except the path D-E-B because the district number E is not used in the production plan.

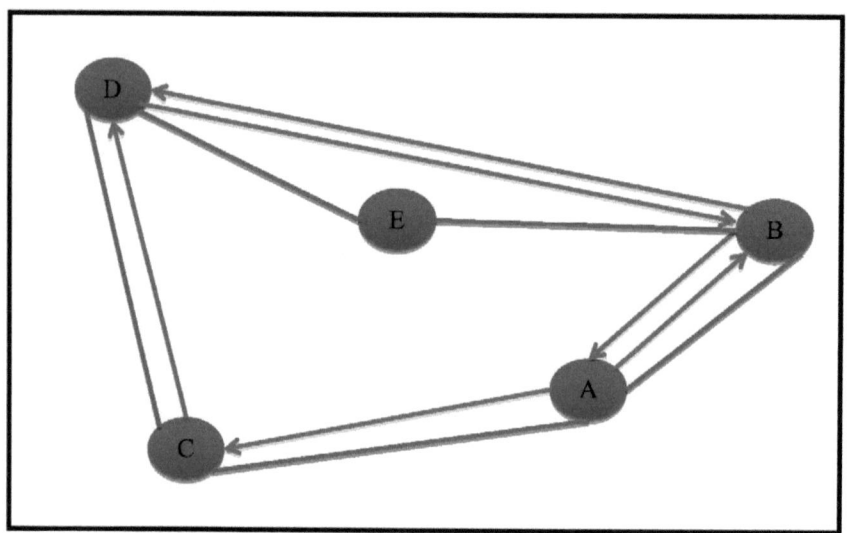

FIGURE 7. 2: PRODUCTION SEQUENCE

As shown in table 7.3, district number D has the largest number of stands to visit in week 4. Thus we have used the developed tabu search algorithm to determine the chipping sequence inside that district in week 4. This is present in figure 7.3 where the starting and ending node are represented in red cycles and the rest of the nodes in blue color. The sequence in this case is the shortest Hamiltonian path because of the small number of stands to cover. This district has a total of 13 stands. Thus, 5 stands are not visited week 4. The line representing the sequence goes through the selected stands to be visited. It can be seen from figure 7.3 that some stands a located very close to each other which is the case in the Danish forest.

46

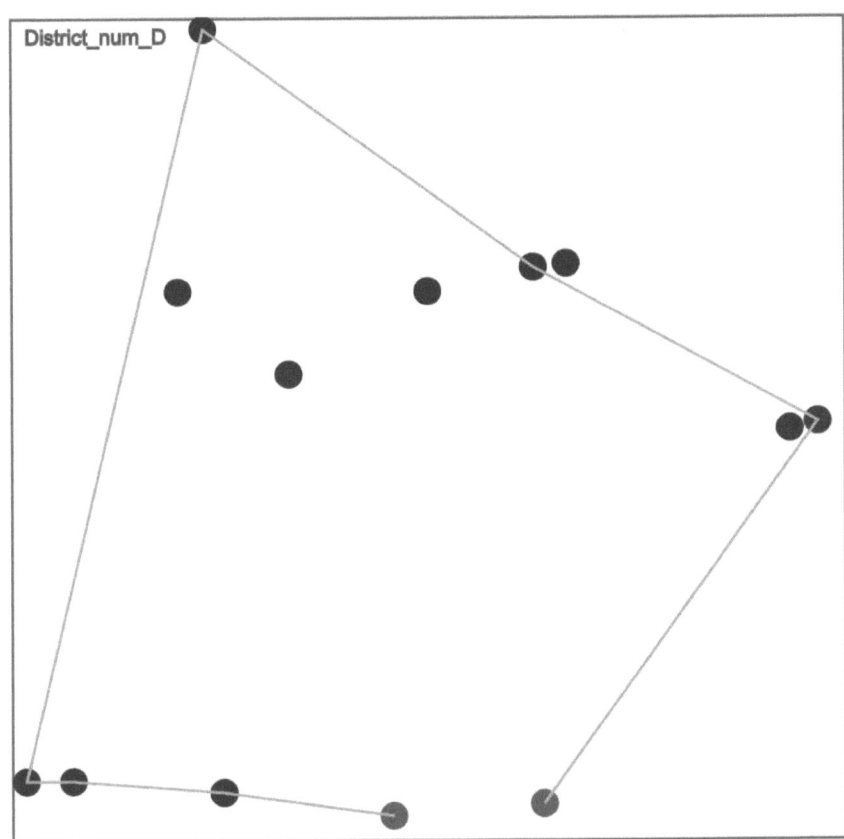

FIGURE 7. 3: INTRA-DISTRICT ROUTING

We have expected the stands with high production rate to be produced in winter time and the stands with low and average production rate in the other seasons. This is not the case because of the low volume of stands with low production rate and the inventory minimization concern (see figure 7.4). If the number of working hours in winter time is reduced, then the stands with high production rate will be selected for the production in winter and the other in low season.

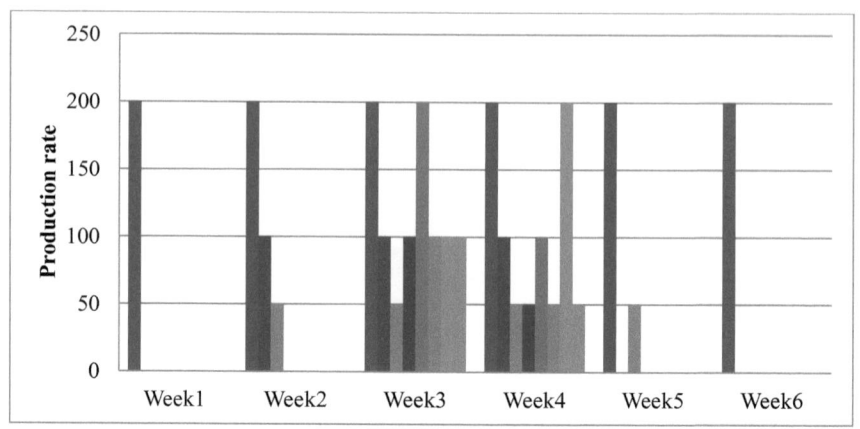

FIGURE 7. 4: INFLUENCE OF WOOD CHIPS QUALITY IN THE PRODUCTION PLAN

Low inventory and high travel costs

We decide now to give more importance to the routing cost between districts. The unit holding cost is then fixed to zero. The production plan is presented in table 7.5. Because of the high routing cost or low inventory cost, the production is not run in week 4 and 5.

Week	1	2	3	4	5	6
Demand (m3)	44	98	271	284	91	42
Stand used	7A, 19A, 21A, 2B, 6B, 10B, 13B, 22B, 25B, 3D, 4D, 14D, 26D, 27D, 28D	8D, 15D, 24D	9D, 11D, 29D			17 D
Produstion volume (m3)	573.36	122.75	97.05	0	0	38. 7
Inventory level (m3)	529.36	554.11	380.16	96.1 6	5.16	1.8 6

TABLE 7. 5: RESULTS FOR LOW INVENTORY COST CASE

Compare to the previous case, the inventory level is very high and the routing cost is very low. In fact, the chipper will only move from district B to A and finally to D (see figure 7.6). The production level in week one is by far higher than that of the demand in the same week. It can be considered as a decreasing function during the time horizon. The demand and production volume follow a totally different pattern (see figure 7.5).

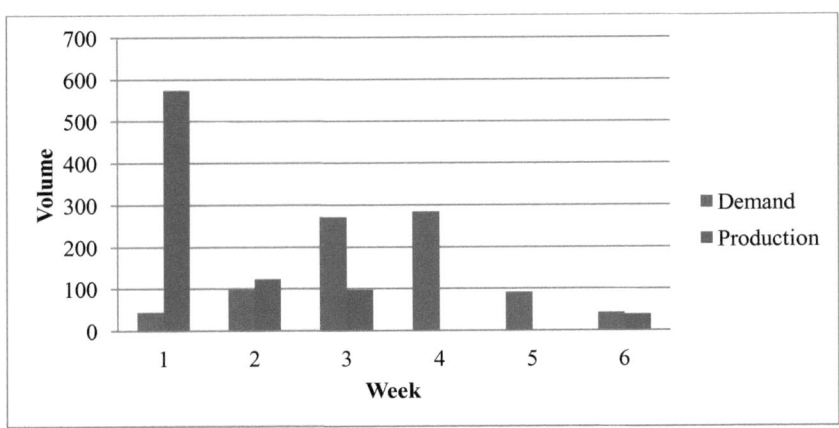

FIGURE 7. 5: PRODUCTION LEVEL

The production schedule is presented in table 7.6 below. From Wednesday at the first week to the last week, the production is run only at district number D. Because of the time and volume capacity restriction, the production is not run in only one district and one week.

Schedue	Week					
District	1	2	3	4	5	6
A	Tuesday 10.30am					
B	Monday 8.00am					
C						
D	Wednesday 8.00am	Monday 8.00am	Monday 8.00am			Monday 8.00am
E						

TABLE 7. 6: PRODUCTION SCHEDULE

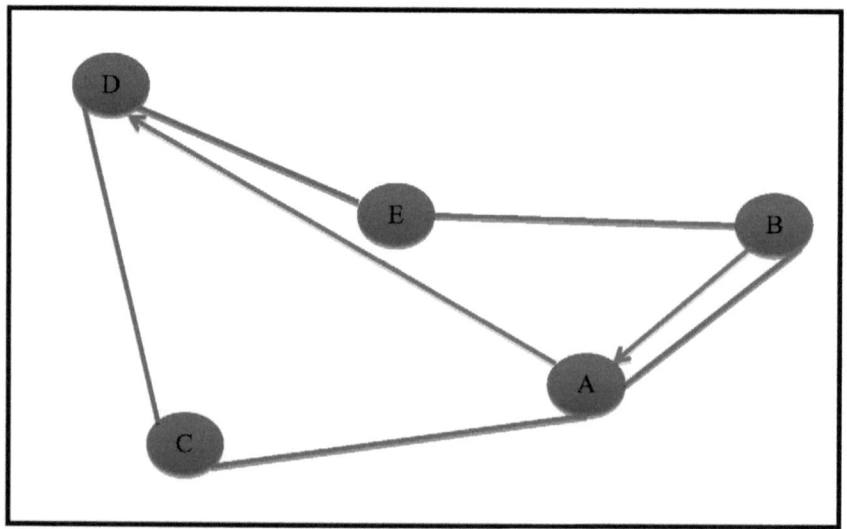

FIGURE 7. 6: ROUTING BETWEEN DISTRICTS

Average inventory unit holding cost

In this case, the unit holding cost is fixed to NOK 10 and the other parameters are unchanged. As can be seen in the production plan presented in table 7.7, the weekly demand and production volume follows the same pattern. That is, the inventory level is very low as expected.

Week	1	2	3	4	5	6
Demand (m3)	44	98	271	284	91	42
Stand used	7A	2B, 6B, 13B	10B, 22B, 25B, 5C, 14D, 30E	1C, 12C, 16C, 23C, 8D, 11D, 15D,	20B, 3D, 24D	18A, 21A
Produstion volume (m3)	44.5	97.54	270.96	284.01	91.05	42.46
Inventory level (m3)	0.5	0.04	0	0.01	0.06	0.52

TABLE 7. 7: PRODUCTION PLAN

Figure 7.7 below shows the level of inventories with high and average unit holding costs. We can easily conclude that there is no significant difference between the inventory level with high unit holding cost and the one with an average unit cost. We have run the model with different unit holding cost and the result were pretty well the same when the unit holding cost was above NOK 10.

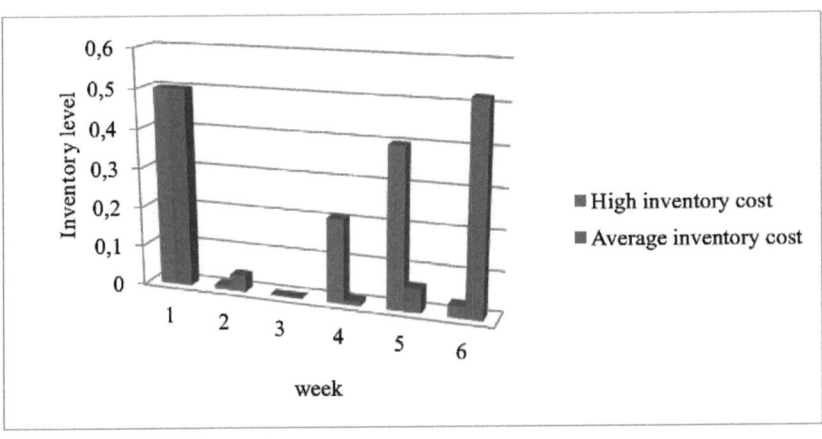

FIGURE 7. 7: INVENTORY LEVEL COMPARISON

The production schedule is presented in table 7.8. in this case, all the districts are visited for the production.

Schedule	Week					
District	1	2	3	4	5	6
A	Monday 8.00am	Monday 8.00am				Monday 8.00am
B			Monday 8.00am		Tuesday 8.00am	
C			Friday 8.00am	Monday 8.00am		
D			Thursday 8.00am	Tuesday 8.00am	Monday 8.00am	
E			Tuesday 8.00am			

TABLE 7. 8: PRODUCTION SCHEDULE

The chipper will start and complete the production at district number A by following the route: A-B-E-D-C-D-B-A (see figure 7.8). The shortest Hamiltonian cycle in presented in red color as in the first case.

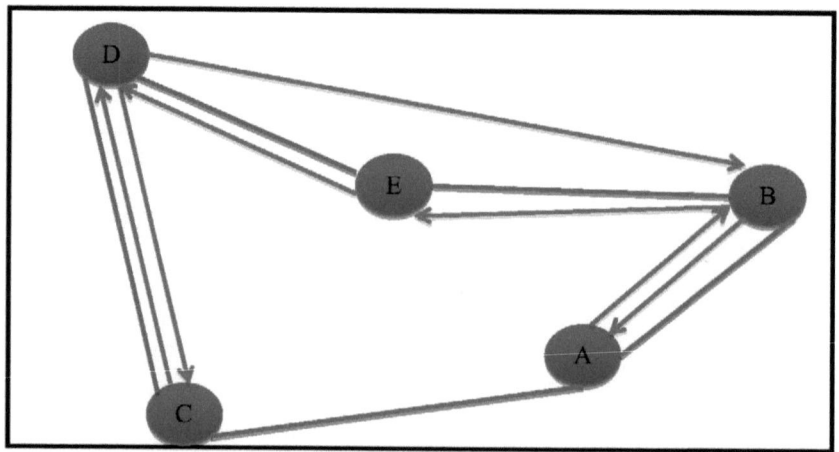

FIGURE 7. 8: PRODUCTION ROUTE

7.1.2 TEST RESULTS AND ANALYSIS FOR LARGE SIZE INSTANCES

While running theses instances, the solver has stopped because of unknown technical problems. The new license has been bought but not installed yet. However, the model will be run with large instances as well as real world instances when the new version will be installed, as a continuation of the project.

7.1.3 TEST RESULTS AND ANALYSIS FOR REAL-WORLD INSTANCES

The test case for real world instances were not used because of the two followings reasons:

1. The total weekly demands and the geographical location of the stands and districts were not provided in due time.
2. The CPLEX version at the university is not working properly and the new order license is not installed yet.

In addition, we expect the problem to be too large for exact methods.

By running the model with real word data, some sensitivity analysis and scenarios will be of great benefit to the supplier of wood chips.

CHAPTER 8 CONCLUSIONS AND FUTURE RESEARCH

Conclusively, a comprehensible description of the wood chips supply chain has been presented, together with a mathematical abstraction of the problem. To assist the supplier in reducing inventory and production cost, a yearly production mix integer programming model has been developed with a 2-opt tabu tearch algorithm for intra-district routing planning. For each week in a year, the model determines the optimal number and specific stands to be used for the production and the starting and ending production time. This result in an optimal inventory level as well.

Several interesting points can be considered for future research:

- To reduce the expected high computing time, the MIP model can be modified to only determine the optimal production capacity at each district and the routing between districts. Then a suitable heuristic algorithm can be used to determine the number and specific stands needed to run the production each week as well as the intra-district routing.

- The developed MIP can be extended to consider possible outsourcing from Finland for example when the demand is larger than the capacity. In addition, constraints on old stands for which the production has to be run can be added. In fact, the production may not be run at some stands for years because of the routing and inventory minimization concern.

- The company is actually using one chipper to run the production. In addition, our model did not consider the influence the production planning on the transportation cost to customers. Thus, a model that considers the location of costumer and the optimal number of machine may be considered. A simulation study may also be included to help the supplier in deciding on preferable and good options.

- An inventory routing planning for customers with a transportation planning from districts to terminal storage may be developed together with the transportation of un-chipped woods to customers.

- The problem might also be approached by considering a stochastic demand and/or inventory replenishments with time windows.
- A storage location problem may be developed to reduce the production and transportation cost. It will be of great benefit to determine the optimal number and location of inventory storages.
- Several Meta heuristic methods such as Unified Tabu Search may be used to solve the problem as well.
- The provided excel shit contain information for the harvest and chipping from 2002 to 2011. An empirical study may be conducted to evaluate the company's performance during the past 10 years.
- Cost benefit analysis study may be conducted to assess the benefit of extending the inventory storage capacity at each customer (plant).
- Some stands have been chipped after being felled for only few weeks. This highly reduces the wood chip quality and increases the holding costs. We suggest a study on the harvesting methods and planning.
- Possible studies may be conducted in Norway and benchmark with Denmark case.
- The available production time is not the same every week. The model can be adjusted to consider the holidays, extra times and limited production time in winter.

REFERENCES

Aghezzaf, El-Houssaine, Birger Raa, and Landeghem Van, Hendrik. 2006. Modeling inventory routing problems in supply chains of high consumption products. *European Journal of Operational Research* 169 (3):1048-1063.

Al-Khayyal, F, and S-J. Hwang. 2007. Inventory constrained maritime routing and

scheduling formulti-commodity liquid bulk, part I: applicationsandmodel. *European Journal of Operational Research* 176.

Amir, Ahmadi Javid, and Nader Azad. 2010. Incorporating location, routing and inventory decisions in supply chain network design. *Transportation Research Part E: Logistics and Transportation Review* 46 (5):582-597.

Andersson, Henrik, Arild Hoff, Marielle Christiansen, Geir Hasle, and Arne Løkketangen. 2010. Industrial aspects and literature survey: Combined inventory management and routing. *Computers & Operations Research* 37 (9):1515-1536.

Anily, Shoshana, and Julien Bramel. 2004. An asymptotic 98.5%-effective lower bound on fixed partition policies for the inventory-routing problem. *Discrete Applied Mathematics* 145 (1):22-39.

Bard, Jonathan F., and Narameth Nananukul. 2009. Heuristics for a multiperiod inventory routing problem with production decisions. *Computers & Industrial Engineering* 57 (3):713-723.

Bard, Jonathan F., and Narameth Nananukul. 2010. A branch-and-price algorithm for an integrated production and inventory routing problem. *Computers & Operations Research* 37 (12):2202-2217.

Bell, WJ, LM Dalberto, ML Fisher, Aj Greenfield, R Jaikumar, and P Kedia. 1983. Improving the distribution of industrial gases with an on-line computerized routing scheduling optimizer. *Interfaces* 13:4-23.

Bengt, Hillring. 2006. World trade in forest products and wood fuel. *Biomass and Bioenergy* 30 (10):815-825.

Bertazzi, Luca, Adamo Bosco, Francesca Guerriero, and Demetrio Laganà. A stochastic inventory routing problem with stock-out. *Transportation Research Part C: Emerging Technologies* (0).

Chen, Yee Ming, and Chun-Ta Lin. 2009. A coordinated approach to hedge the risks in stochastic inventory-routing problem. *Computers & Industrial Engineering* 56 (3):1095-1112.

Christiansen, Marielle. 1999. Decomposition of a Combined Inventory and Time Constrained Ship Routing Problem. *Transportation Science* 33 (1):3-26.

Christopher, Martin. 2011. *Logistics & supply chain management*. London: Financial Times Prentice Hall.

Darwish, M. A., and O. M. Odah. 2010. Vendor managed inventory model for single-vendor multi-retailer supply chains. *European Journal of Operational Research* 204 (3):473-484.

Dimitrijević, Vladimir, and Zoran Šarić. 1997. An efficient transformation of the generalized traveling salesman problem into the traveling salesman problem on digraphs. *Information Sciences* 102 (1–4):105-110.

Disney, S. M., A. T. Potter, and B. M. Gardner. 2003. The impact of vendor managed inventory on transport operations. *Transportation Research Part E: Logistics and Transportation Review* 39 (5):363-380.

Dong, Yan, and Kefeng Xu. 2002. A supply chain model of vendor managed inventory. *Transportation Research Part E: Logistics and Transportation Review* 38 (2):75-95.

Gendreau, Michel, Gilbert Laporte, and Frédéric Semet. 1998. A tabu search heuristic for the undirected selective travelling salesman problem. *European Journal of Operational Research* 106 (2–3):539-545.

Harrison, Alan, and Remko van Hoek. 2011. *Logistics management and strategy: competing through the supply chain*. Harlow: Financial Times Prentice Hall.

Helle, Serup, Falster Hans, Gamborg Christian, Gundersen Per, Hansen Leif, Heding Niels, H. Jakobsen Henrik, Kofman Pieter, Nikolaisen Lars, and M. Thomsen Iben. 2002. *Wood for Energy Production*

Technology - Environment - Economy. Edited by S. Helle. 2 ed.

Herer, Yale T., and Roberto Levy. 1997. The Metered Inventory Routing Problem, an integrative heuristic algorithm. *International Journal of Production Economics* 51 (1-2):69-81.

Huang, Shan-Huen, and Pei-Chun Lin. 2010. A modified ant colony optimization algorithm for multi-item inventory routing problems with demand uncertainty. *Transportation Research Part E: Logistics and Transportation Review* 46 (5):598-611.

Hugos, Michael. 2011. *Essentials of supply chain management.* Hoboken, N.J.: Wiley.

J-F, Cordeau, G Laporte, and A Mercier. 2001. A unified tabu search heuristic for vehicle routing

problems with time windows. *Journal of the Operational Research Society* 52:928–936.

Kalle, Kärhä. 2011. Industrial supply chains and production machinery of forest chips in Finland. *Biomass and Bioenergy* 35 (8):3404-3413.

Karapetyan, D., and G. Gutin. 2011. Lin–Kernighan heuristic adaptations for the generalized traveling salesman problem. *European Journal of Operational Research* 208 (3):221-232.

Karapetyan, D., and G. Gutin. 2012. Efficient local search algorithms for known and new neighborhoods for the generalized traveling salesman problem. *European Journal of Operational Research* 219 (2):234-251.

Kleywegt, A. J, V. S Nori, and M. W. P Savelsbergh. 2002. The Stochastic Inventory Routing Problem with

Direct Deliveries. *Transportation Science* 36:94-118.

Larson, Richard C. 1988. Transporting sludge to the 106-mile site: An inventory/routing model for fleet sizing and logistics system design. *Transportation Science* 22 (3):186–198.

Lauri, Pekka, A. Maarit I. Kallio, and Uwe A. Schneider. 2012. Price of CO2 emissions and use of wood in Europe. *Forest Policy and Economics* 15 (0):123-131.

Lei, Lei, Liu Shuguang, Ruszczynski Andrzej, and Park Sunju. 2006. On the integrated production,

inventory, and distribution routing problem. *IIE Transactions on Scheduling &*

Logistics 38 (11):955–970.

Li, Jianxiang, Haoxun Chen, and Feng Chu. 2010. Performance evaluation of distribution strategies for the inventory routing problem. *European Journal of Operational Research* 202 (2):412-419.

Liu, Shu-Chu, and Jyun-Ruei Chen. 2011. A heuristic method for the inventory routing and pricing problem in a supply chain. *Expert Systems with Applications* 38 (3):1447-1456.

Liu, Shu-Chu, and Wei-Ting Lee. 2011. A heuristic method for the inventory routing problem with time windows. *Expert Systems with Applications* 38 (10):13223-13231.

Mizaraitė, D., S. Mizaras, and L. Sadauskienė. 2007. Wood fuel supply, costs and home consumption in Lithuania. *Biomass and Bioenergy* 31 (10):739-746.

Moin, N. H., S. Salhi, and N. A. B. Aziz. 2011. An efficient hybrid genetic algorithm for the multi-product multi-period inventory routing problem. *International Journal of Production Economics* 133 (1):334-343.

Möller, Bernd, and Per S. Nielsen. 2007. Analysing transport costs of Danish forest wood chip resources by means of continuous cost surfaces. *Biomass and Bioenergy* 31 (5):291-298.

Pentti, Hakkila. 2006. Factors driving the development of forest energy in Finland. *Biomass and Bioenergy* 30 (4):281-288.

Popović, Dražen, Nenad Bjelić, and Gordana Radivojević. 2011. Simulation Approach to Analyse Deterministic IRP Solution of the Stochastic Fuel Delivery Problem. *Procedia - Social and Behavioral Sciences* 20 (0):273-282.

R, Jirjis. 1995. Storage and drying of wood fuel. *Biomass and Bioenergy* 9 (1–5):181-190.

Savelsbergh, Martin, and Jin-Hwa Song. 2007. Inventory routing with continuous moves. *Computers & Operations Research* 34 (6):1744-1763.

Savelsbergh, Martin, and Jin-Hwa Song. 2008. An optimization algorithm for the inventory routing problem with continuous moves. *Computers & Operations Research* 35 (7):2266-2282.

Shi, X. H., Y. C. Liang, H. P. Lee, C. Lu, and Q. X. Wang. 2007. Particle swarm optimization-based algorithms for TSP and generalized TSP. *Information Processing Letters* 103 (5):169-176.

Tahvanainen, Timo, and Perttu Anttila. 2011. Supply chain cost analysis of long-distance transportation of energy wood in Finland. *Biomass and Bioenergy* 35 (8):3360-3375.

Tsubakitani, Shigeru, and James R. Evans. 1998. An empirical study of a new metaheuristic for the traveling salesman problem. *European Journal of Operational Research* 104 (1):113-128.

Tsubakitani, Shigeru, and James R. Evans. 1998. Optimizing tabu list size for the traveling salesman problem. *Computers & Operations Research* 25 (2):91-97.

Tyan, Jonah, and Hui-Ming Wee. 2003. Vendor managed inventory: a survey of the Taiwanese grocery industry. *Journal of Purchasing and Supply Management* 9 (1):11-18.

Voudouris, Christos, and Edward Tsang. 1999. Guided local search and its application to the traveling salesman problem. *European Journal of Operational Research* 113 (2):469-499.

Yang, Jinhui, Xiaohu Shi, Maurizio Marchese, and Yanchun Liang. 2008. An ant colony optimization method for generalized TSP problem. *Progress in Natural Science* 18 (11):1417-1422.

Yu, Yugang, Haoxun Chen, and Feng Chu. 2008. A new model and hybrid approach for large scale inventory routing problems. *European Journal of Operational Research* 189 (3):1022-1040.

TABLE 1: AVAILABLE PRODUCTION VOLUME FOR EACH DISTRICT FROM MAY 2010 TO MAY 2011

District	Number of Stands	Volume (m³)
1007	6	1191.00
1302	103	6615.00
1307	98	13654.00
1807	25	1777.00
2007	54	3704.00
2107	2	268
3007	150	9459.15
3107	64	8625.65
3204	10	1409.00
3207	56	3608.00
4407	69	6800.00
5307	106	17727.00
5501	4	2450.00
5507	283	39285.00
5707	94	13120.00
6007	90	19127.00
6207	267	40315.00
6507	136	11060.00
6607	61	7084.00
7007	29	1882.00
7207	25	2777.00

The production was run in 28 weeks from May 2010 to may 2011 with a total of 21 districts and 1734 stands available (see table 1).

For a given week, the chipping is run at stands from different districts. That means many districts are visited more than once for production.

TABLE 2: MEAN PRODUCTION RATE FOR EACH TREE SPECIE

		Production Rate (m³/hour)
		Mean
Tree Species		411.90
	ægr	.
	ÆGR	487.91
	ÆR	525.54
	agr	.
	AGR	460.47
	ALØ	436.97
	ANÅ	441.50
	ASK	348.08
	ASP	.
	BIR	437.50
	bjf	.
	BJF	461.28
	bøg	.
	BØG	588.28
	cof	.
	COF	401.83
	CYP	.
	dgr	.
	DGR	486.32
	DIV	.
	EG	435.79
	EL	560.00
	ELM	.
	fbf	.
	FBF	547.61
	FYR	600.00
	hgr	.

	HGR	375.00
	KIR	.
	lær	.
	LÆR	502.47
	LIN	342.11
	LØV	.
	NÅL	423.05
	ngr	.
	NGR	450.74
	nob	.
	NOB	517.65
	omo	.
	OMO	499.98
	ØSF	533.30
	POP	.
	REG	403.24
	REL	800.00
	rgr	.
	RGR	517.02
	sgr	.
	SGR	523.59
	skf	.
	SKF	524.72
	THU	.
	TSU	.
	WEY	.

Base on the information given in the excel file, some tree species have not been harvested yet, which means it is impossible to have the production rate. The missing values are represented by a dot in table 2. Figure 1 below shows the frequency of the different wood quality available in the planning period of May 2010 – May 201.

FIGURE 1: TREE SPECIES

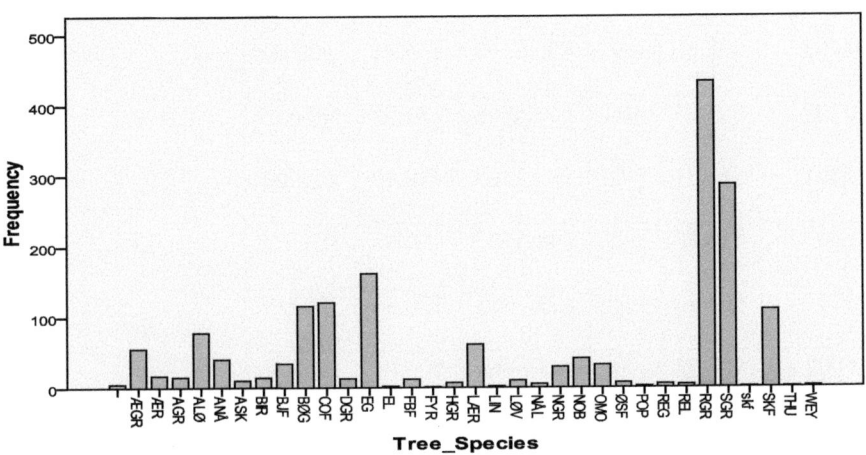

TABLE 3: MEAN PRODUCTION RATE BETWEEN DISTRICTS

Report					
Production_Rate					
District_Num	Mean	N	Std. Deviation	Minimum	Maximum
1007	511.6729	6	211.51247	383.87	900.00
1302	449.5459	102	246.56798	57.14	1964.29
1307	481.7991	88	243.87758	97.27	1960.78
1807	460.8192	25	326.32812	43.24	1250.00
2007	519.2219	54	198.81293	59.09	1200.00
2107	174.6753	2	15.61145	163.64	185.71
3007	484.3768	144	483.63950	57.14	5400.00
3107	578.6205	64	715.63823	54.55	6000.00

3204	541.0027	10	380.48245	150.00	1200.00
3207	532.2150	56	692.31060	37.84	5500.00
4407	398.6029	69	85.16551	105.48	500.00
5307	503.2734	104	565.59862	131.58	5666.67
5501	542.7193	4	98.63307	419.16	625.00
5507	454.8042	275	248.57354	42.86	3833.33
5707	661.4319	94	473.73528	136.36	3200.00
6007	459.5386	90	241.86612	106.06	1954.55
6207	468.8633	267	212.92764	71.43	2100.00
6507	532.7436	136	336.23549	111.11	2000.00
6607	486.4396	59	316.54122	32.14	2545.83
7007	410.5930	29	257.06971	52.63	1200.00
7207	496.2486	25	51.61283	450.00	625.00
Total	490.0164	1703	364.81181	32.14	6000.00

TABLE 4: TEST OF MEANS DIFFERENCE BETWEEN DISTRICTS

Case Processing Summary							
	Cases						
	Included		Excluded		Total		
	N	Percent	N	Percent	N	Percent	
Production_Rate * District_Num	1703	98.2%	31	1.8%	1734	100.0%	

ANOVA Table							
			Sum of Squares	df	Mean Square	F	Sig.
Production_Rat e * District_Num	Between Groups	(Combin ed)	5419807 .538	20	270990. 377	2.062	.004
	Within Groups		2.211E8	1682	131447. 911		
	Total		2.265E8	1702			

TABLE 5: TEST OF MEANS DIFFERENCE BETWEEN DISTRICTS

From Table 5, we can see that there is a statistical difference between means production rate and districts. The table that shows exactly which means production rates are statistically different is not included here because of the large size of the table.

Tree_Diameter					
		Frequen cy	Percen t	Valid Percent	Cumulativ e Percent
Valid	7.00	7	1.0	1.1	1.1
	8.00	13	1.9	2.0	3.1
	9.00	6	.9	.9	4.0
	10.00	73	10.8	11.2	15.2
	11.00	78	11.6	12.0	27.2
	12.00	58	8.6	8.9	36.1
	13.00	35	5.2	5.4	41.5
	14.00	36	5.3	5.5	47.0
	15.00	153	22.7	23.5	70.5
	16.00	40	5.9	6.1	76.7
	17.00	11	1.6	1.7	78.3
	18.00	21	3.1	3.2	81.6
	19.00	9	1.3	1.4	82.9
	20.00	51	7.6	7.8	90.8
	21.00	6	.9	.9	91.7
	23.00	1	.1	.2	91.9
	24.00	5	.7	.8	92.6
	25.00	41	6.1	6.3	98.9

	26.00	1	.1	.2	99.1
	27.00	1	.1	.2	99.2
	30.00	5	.7	.8	100.0
	Total	651	96.6	100.0	
Missing	System	23	3.4		
Total		674	100.0		

Printed by Books on Demand GmbH, Norderstedt / Germany